D1090595

Space Stations: Living and Working in Space

Amanda Davis

The Rosen Publishing Group's
PowerKids Press™
New York

Published in 1997 by The Rosen Publishing Group, Inc.
29 East 21st Street, New York, NY 10010

First Edition

Book Design: Erin McKenna

Photo Credits: Cover and pp. 4, 20 © Jack Zehrt/FPG International Corp.; pp. 7, 11, 12 © Archive Photos; p. 8 © NASA 1988; p. 12 © 1989 NASA/SB; p. 16 © Jeff Kaufman 1993/FPG International Corp.; p. 19 © FPG International Corp.

Davis, Amanda
Space stations: living and working in space / by Amanda Davis
p. cm. — (Exploring Space)
Includes index.
Summary: Discusses what space stations are, why they are important, what kinds of research is done on them, and the international cooperation they foster.
ISBN 0-8239-5062-X
1. Space stations—Juvenile literature. [1. Space stations.]
I. Title. II. Series: Davis, Amanda. Exploring space.
TL797.D38 1997
629.44'2—dc21

96-54497
CIP
AC

Manufactured in the United States of America

Contents

What Is a Space Station?

Have you ever wondered what it would be like to live in space? Just read about **astronauts** (AS-tro-nots) and **scientists** (SY-en-tists) who have lived and worked on a **space station** (SPAYS STAY-shun).

A space station is a **laboratory** (LAB-rah-TOR-ee) in space that **orbits** (OR-bits) Earth. Scientists and astronauts live in these stations for months at a time. They do amazing **experiments** (ex-PEER-uh-ments) that help them learn a lot about the way things work on Earth and in space.

Scientists can see and study Earth in a different way from space.

SPACE FACT

When Skylab fell, pieces of it landed in Australia.

Skylab

The first American space station, Skylab, was **launched** (LAWNCHT) into space in 1973.

Three different groups of astronauts lived and worked on Skylab. One group stayed for almost three months. The astronauts traveled to and from Skylab in spaceships like the ones used to take people to the moon.

But Skylab didn't stay in space forever. In 1979, Skylab fell back to Earth and was destroyed.

The last group of people left Skylab in 1974. ▶

Mir

In 1981, Russia launched the first **permanent** (PER-muh-nent) space station into orbit. It is called **Mir** (MEER). Mir is the Russian word for "peace."

The space station Mir is very big. It is 107 feet long and 90 feet wide. It orbits 240 miles above Earth. Mir has traveled more than 1.5 billion miles since it was launched into space.

In 1995, American **space shuttles** (SPAYS SHUT-uhlz) brought astronauts and **supplies** (suh-PLYZ) to Mir. Scientists and astronauts from Japan, England, Australia, and many other countries have lived and done **research** (REE-serch) aboard Mir.

◀ Spaceships are used to carry supplies and astronauts to and from Mir.

Shannon Lucid

Astronaut Shannon Lucid did something that no one has ever done before. She stayed in space longer than any other American in history. She lived on the Mir space station for six months, from March to September 1996.

Shannon did many experiments and worked side by side with Russian astronauts. She even sent letters by **e-mail** (E-mayl) to her family every day from space!

When she got back to Earth, President Clinton gave Shannon the Congressional Space Medal for her great work as an astronaut and scientist.

Shannon and Russian astronaut Yuri Usachev get ready to eat a meal on Mir.

Experiments

Space lets scientists study how things act and move without **gravity** (GRAV-ih-tee). This natural **force** (FORSS) causes objects to be attracted to each other. Without gravity, everything, including people, would float around in the air!

Studying these things in space without gravity lets scientists look at how things work in a different way.

Because of astronauts and the experiments they do in space, we've learned more about the effects of exercise on the human body. We've even learned new ways to make things that people use every day, such as paint and **contact lenses** (KON-takt LEN-zez).

◀ Being in a space station lets scientists study the effects of the absence of gravity.

Living in Space

Gravity affects everything we do on Earth. The way we eat, sleep, and move would be very different without gravity.

When we sleep, gravity keeps us from floating out of our beds. In space, astronauts float while they sleep because there is no gravity. Some attach themselves to a wall so they don't float. Some sleep upside down or sideways. Without gravity, it doesn't matter how an astronaut sleeps. In space, it feels the same whichever way you sleep!

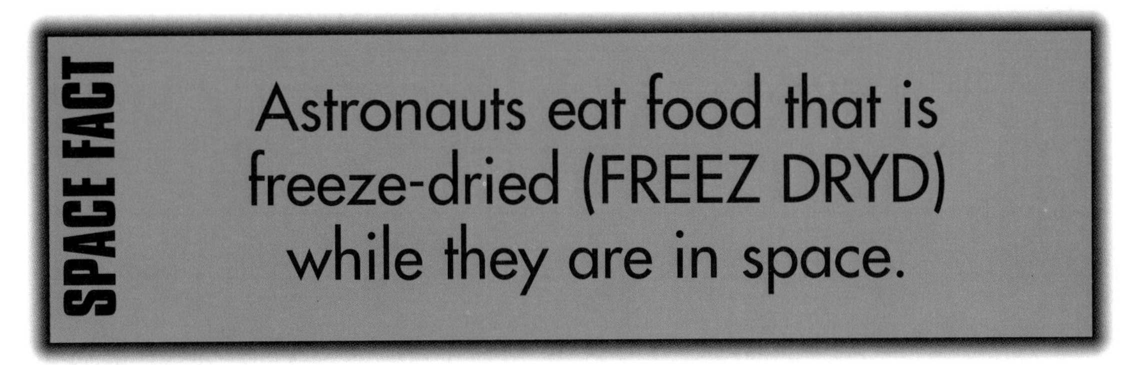

Astronauts must learn to get used to living without gravity. ▶

Bathing

Astronauts that are living on a space station can't take baths or showers like we can. The water would float around the spaceship in little blobs instead of staying in the tub like it does on Earth!

So astronauts use a small water hose and a washcloth. They rub the washcloth with a little soap and use it with the water hose to wash their bodies. That's the closest thing to a bath that astronauts have.

◀ Gravity keeps the water in the bathtub when we bathe. Astronauts aren't so lucky!

The International Space Station

Right now, a worldwide project is underway to build the biggest space station in history. Thirteen countries are planning to build the International Space Station. Parts of the station are being built in Japan, Canada, the United States, Russia, and other European countries.

The pieces of the station are being put together in space. In 2002, after 44 flights into space, the station will be finished and ready for its first astronauts.

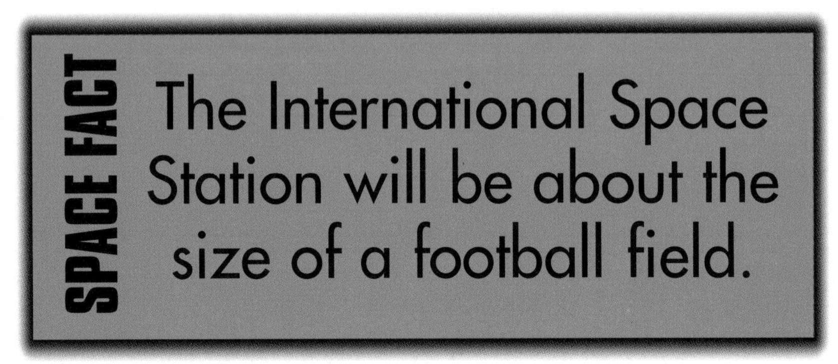

Artists have often painted pictures of space stations to help scientists imagine what they will look like. ▶

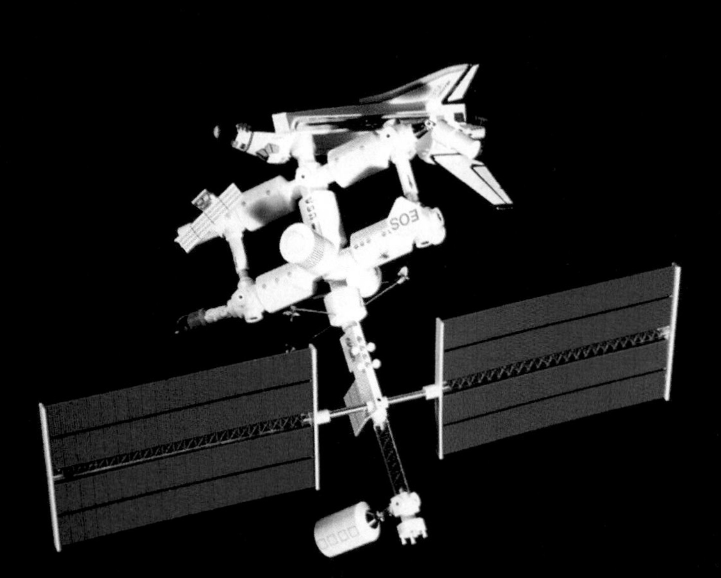

What Will the Space Station Be Like?

When the space station is completed, it will orbit around Earth 250 miles up in the sky. Regular airplanes hardly ever go higher than seven miles above Earth's surface.

There will be seven laboratories inside the station for scientific experiments. Six people at a time will be able to live and work on the station.

Space shuttles will carry new supplies and astronauts to and from the space station.

◀ A space shuttle will attach itself to the space station when delivering supplies.

Nations Working Together

The International Space Station will teach us a lot about science. With good luck and hard work, it may even lead to people living in space.

One of the most important things that the space station will do for our planet is to bring people together. Thirteen nations are working on one project. They are helping each other to reach the same goal. The space station shows us how much we can do when we work together.

Glossary

astronaut (AS-tro-not) A person who travels in space.

contact lenses (KON-takt LEN-zez) Small, thin pieces of plastic that are put on the eyes to improve a person's vision.

e-mail (E-mayl) Electronic mail.

experiment (ex-PEER-uh-ment) A test done on something to learn more about it.

force (FORSS) Something in nature that causes action.

freeze-dried (FREEZ DRYD) When food is dried by freezing and removing the water and air.

gravity (GRAV-ih-tee) A natural force that causes objects to be attracted to each other.

laboratory (LAB-rah-TOR-ee) A room where scientists use special equipment to do experiments.

launch (LAWNCH) To lift off.

Mir (MEER) The name of the Russian space station. In Russian, it means "peace."

orbit (OR-bit) How one thing circles another.

permanent (PER-muh-nent) Meant to last for a very long time.

research (REE-serch) Studying and finding out facts about something.

scientist (SY-en-tist) A person who studies the world and the universe.

space shuttle (SPAYS SHUT-uhl) A special spaceship that can be used many times, and lands like an airplane.

space station (SPAYS STAY-shun) A structure in space where scientists and astronauts live and work for long periods of time.

supplies (suh-PLYZ) Things a person needs to do a job.

Index